Learning About
Acid Rain

*A Teacher's Guide
For Grades 6
Through 8*

EPA
United States
Environmental Protection
Agency

CLEAN AIR
MARKET PROGRAMS

United States Environmental Protection Agency

Office of Air and Radiation

Office of Atmospheric Programs

Clean Air Markets Division (6204J)

1200 Pennsylvania Ave., NW

Washington, D.C. 20460

EPA 430-F-08-002

April 2008

A History
of the
Acid Rain Program

1970 – 1994

1970
Twenty million people celebrate the first Earth Day.

1970
The Clean Air Act (CAA) is passed.

1977
Congress strengthens the CAA and includes requirements for SO_2 pollution control at power plants.

1978
The National Atmospheric Deposition Program/National Trends Network (NADP/NTN) begins monitoring sulfur and nitrogen deposition to ecosystems.

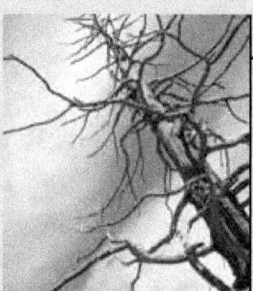

1980
The National Acid Precipitation Assessment Program (NAPAP), mandated by Congress, begins study on acid rain.

1980
Lake acidification and fish loss in the Adirondacks, Green Mountains, and Sierra Nevada make national news.

1986
The United States and Canada begin study of cross-border acid rain transport. The United States is called upon to reduce emissions of SO_2 and NO_X, especially from coal-burning power plants.

1987
The Clean Air Status and Trends Network (CASTNET) is established to monitor dry deposition.

1990
Congress strengthens the CAA and establishes the Acid Rain Program using a market-based approach to reduce SO_2 from power plants by more than 50 percent.

1993
EPA publishes acid rain regulations, and the Chicago Board of Trade holds first auction of SO_2 allowances.

1994
Projected costs of compliance re-estimated by the Government Accountability Office and the Electric Power Research Institute at less than half of original estimates.

1995

Phase I of Acid Rain Program implementation begins. SO₂ emissions fall to 5 million tons below 1980 levels. Acidity of rainfall in the eastern United States drops 10 to 25 percent.

2000

Phase II of Acid Rain Program begins, regulating additional smaller/cleaner plants and requiring further reductions in NOₓ and SO₂.

1996

About 150 of the largest coal-fired power plants begin to implement Acid Rain Program NOₓ requirements.

2001

Introduction of the On-line Allowance Tracking System begins an era of paperless allowance transfer recording.

2002

EPA begins electronic audit process to supplement existing rigorous monitoring program.

1997

More than 80 percent of affected companies have engaged in private allowance transactions.

2003

Lakes and streams in the Adirondacks, Upper Midwest, and Northern Appalachian Plateau show signs of recovery.

2004

Acid Rain Program sources emit 34 percent less SO₂ and 43 percent less NOₓ than in 1990, despite a 34 percent increase in fuel usage.

1998

Regulatory revisions enhance efficiencies of compliance and administration. Nearly 10 million economically significant allowance transfers take place.

1999

Allowance banking peaks. SO₂ early reductions total over 11 million tons.

2005

New study estimates 2010 annual Acid Rain Program benefits at $122 billion and annual costs at $3 billion. According to the 2005 NAPAP report, further emission reductions are necessary to achieve broader environmental recovery. EPA promulgates Clean Air Rules to further reduce SO₂, NOₓ, and for the first time, mercury.

Contents

Purpose

Acid rain is a complex environmental problem which affects the United States and many other countries around the world. The United States Environmental Protection Agency (EPA) was established in 1970 to address environmental issues, such as acid rain. Through its programs, EPA works to protect human health and the environment in the United States by developing and enforcing regulations and studying environmental conditions.

In addition, EPA is committed to informing the public about environmental topics and its efforts to solve them through written materials and its Web site (www.epa.gov). EPA frequently receives requests for information about environmental problems from school systems, teachers, and individuals. Acid rain is one of the most frequently requested topics. As part of EPA's public outreach on acid rain, EPA first developed this guide in 1990. This revised guide is designed to help students better understand the science, cause and effect, and regulatory and citizen action that are part of understanding and addressing acid rain.

This book is intended for teachers of students in 6th-8th grade. It is written at a 6th grade level and the language, concepts, and experiments may need to be adapted for other grades accordingly. After reading the guide and doing some of the experiments and activities, we hope that you and your students will have a better understanding of acid rain and the problems it causes, as well as a greater interest in its resolution and in applied environmental science.

In addition to this teacher's guide, EPA has many other publications with information on research, monitoring, regulation, and other aspects of the acid rain problem. If you are interested in learning more, resources are available at www.epa.gov/airmarkets. This guide, "Learning About Acid Rain: A Teacher's Guide for Grades 6 Through 8," is available online at www.epa.gov/acidrain /education/teachersguide.pdf. Printed copies are available for free through the Acid Rain Hotline (202-343-9620).

Introduction

When harmful substances are released into the air, it causes pollution.

Humans make use of many things found in nature. For example, we use trees to build our homes and cotton to make our clothes. Things that are not made by people, but instead occur naturally, are called NATURAL RESOURCES. Some examples of natural resources are plants, minerals, and water. All of these things are important to humans because they provide us with the materials we need to make the things we use everyday. Some of the products made from natural resources are obvious to us, like the timber and stone that make buildings. Other natural resources are not as noticeable, like the underground water table where our drinking water comes from. Natural resources that humans use to generate electricity are called ENERGY RESOURCES. Most energy in the United States comes from burning FOSSIL FUELS such as coal, oil, and natural gas. Coal, oil, and natural gas are called fossil fuels because they were formed millions of years ago from dead plants and animals.

People burn fossil fuels for many reasons. We burn oil and coal to make the electricity that we need to light buildings and run appliances like televisions and computers. We burn gas to heat our homes and to power cars, buses, and airplanes. Many human activities, including the burning of fossil fuels, cause POLLUTION. Pollution is the release of harmful substances called POLLUTANTS into the ENVIRONMENT. The air pollution created when fossil fuels burn does not stay in the air forever. Instead it can travel great distances, and fall to the ground again as dust or rain. When airborne chemicals and pollutants fall to the Earth, or deposit, it is called DEPOSITION.

ACID RAIN forms when clean rain comes into contact with pollutants in the air, like SULFUR DIOXIDE (SO_2), CARBON DIOXIDE (CO_2), and NITROGEN OXIDES (NO_x). Although sulfur dioxide and carbon dioxide occur in the air naturally, burning fossil fuels adds more of these chemicals to the air. When these pollutants are released into the air, they mix and react with water, oxygen, and other chemicals to form acid rain. Acid rain then falls to the Earth where it can damage plants, animals, soil, water, and building materials.

TERMINOLOGY

Natural Resources All the parts of the Earth that are not human-made and which people use, like fish, trees, minerals, lakes, or rivers.

Energy Resources Natural resources that can be used to make heat, electricity, or any other form of energy. The most commonly used energy resources are fossil fuels (coal, oil, and gas), but the sun, wind, and anything else that makes energy are also energy resources.

Fossil Fuels Oil, natural gas, and coal. Fossil fuels were made in nature from ancient plants and animals, and today we burn them to make energy.

Pollution The release of harmful substances into the environment.

Pollutants Chemicals or other substances that are harmful to or unwanted in the environment. Some examples of pollutants are sulfur dioxide (SO_2), nitrogen oxides (NO_x), ozone, and particulate matter.

Environment The air, water, soil, minerals, organisms, and all other factors surrounding and affecting an organism.

Deposition When chemicals like acids or bases fall to the Earth's surface. Deposition can be wet (rain, sleet, snow, fog) or dry (gases, particles).

Acid Rain Rain that has become acidic by contact with air pollution. Other forms of precipitation, such as snow and fog, are also often included in the term acid rain or acid wet deposition.

Sulfur Dioxide (SO_2) A naturally occurring gas made of sulfur and oxygen that is also released when fossil fuels are burned.

Carbon Dioxide (CO_2) A naturally occurring gas made of carbon and oxygen. Sources of carbon dioxide in the atmosphere include animals, which exhale carbon dioxide, and the burning of fossil fuels and biomass.

Nitrogen Oxides (NO_x) A family of gases made up of nitrogen and oxygen commonly released by burning fossil fuels.

Particulate Matter Tiny solid particles or liquid droplets suspended in the air.

Ozone A chemical that is made of three oxygen atoms joined together, and found in the Earth's atmosphere. There are two kinds of ozone: good ozone, and bad ozone. Good ozone is found high in the Earth's atmosphere, and prevents the sun's harmful rays from reaching the Earth. Bad ozone is found low to the ground, and can be harmful to animals and humans because it damages our lungs, sometimes making it difficult to breathe.

Emissions The gases that are released when fossil fuels are burned.

Ozone Layer The layer of ozone that shields the Earth from the sun's harmful rays.

Greenhouse Gases Gases that occur naturally in the Earth's atmosphere and trap heat to keep the planet warm. Some examples are carbon dioxide, water vapor, halogenated fluorocarbons, methane, hydrofluorocarbons, nitrous oxide, perfluoronated carbons, and ozone. Some human actions, like the burning of fossil fuels, also produce greenhouse gases.

Despite its name, acid rain does not burn and cannot directly harm people. However, the pollutants that cause acid rain, especially SO_2 and NO_x, can react with other pollutants in the air, forming substances like PARTICULATE MATTER and ground level OZONE, which can sometimes make people sick.

While the "Acid Rain Teacher's Guide" focuses mainly on the issue of acid rain, the EMISSIONS that result from the burning of fossil fuels have many other environmental consequences in addition to causing acid rain. Chemicals like NO_x, produced by the burning of fossil fuels, combine with other chemicals in the atmosphere to form ground level ozone. Although the planet needs an OZONE LAYER for protection from the sun's harmful ultraviolet rays, ozone can be dangerous when it forms low to the ground because it hurts our lungs and sometimes makes it difficult to breathe.

Other chemicals that are released by the burning of fossil fuels are GREENHOUSE GASES. Greenhouse gases occur naturally in the Earth's atmosphere, keeping the planet warm enough for humans to live. Without greenhouse gases, the planet would be an average 60° F colder than it is today (brrr!). However, since the Industrial Revolution, human activity, such as the burning of fossil fuels, has increased the amount of greenhouse gases in the atmosphere. By increasing the levels of greenhouse gases, human activities are affecting the mix of gases in the atmosphere. This is causing the Earth's temperature to rise. For more information on climate change and its causes and effects, check out www.epa.gov/climatechange.

The consequences of air pollution are important to understand because air pollution can be carried long distances and affect large areas. This means that pollution from a town hundreds of miles away may be affecting your community. Scientists, engineers, and researchers monitor the effects of pollution on the air, forests, water, and soil. They are inventing ways to reduce the amount of pollution that enters the environment and to prevent new damage in the future.

Where Our Electricity Comes From

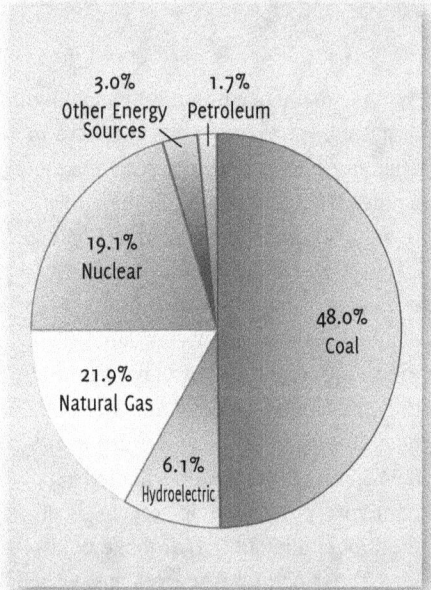

Half of our electricity in the United States is made by coal power plants.

Observations About Acidity

ACIDIC and BASIC are two extremes that describe chemicals, just as hot and cold are two extremes that describe temperature. Mixing ACIDS and BASES can cancel out their extreme effects; much like mixing hot and cold water can even out the water temperature. A substance that is neither acidic nor basic is NEUTRAL. The pH SCALE measures the acidic or basic level of a substance. The pH scale ranges from 0 to 14. A pH of 7 is neutral, while a pH less than 7 is acidic and a pH greater than 7 is basic.

Pure water is neutral. However, when chemicals are mixed with water, the mixture can become either acidic or basic. Examples of acidic substances are vinegar and lemon juice. Laundry detergents and ammonia are examples of basic substances. Chemicals that are very basic or very acidic usually change or alter whatever they meet. Substances that have this property are called REACTIVE. You should be careful with these kinds of chemicals because they can cause severe burns and are often toxic if swallowed. For example, household drain cleaners often contain lye, a very basic chemical that is reactive and could burn you.

TERMINOLOGY

Acidic Describes a substance with a pH less than 7.

Basic Describes a substance with a pH greater than 7. Another word for basic is alkaline.

Acid Any of a large group of chemicals with a pH less than 7. Examples are battery acid, lemon juice, and vinegar.

Base Any of a large group of chemicals with a pH greater than 7. Examples are ammonia and baking soda.

Neutral A substance that is neither an acid nor a base and has a pH of 7. Neutral substances can be created by combining acids and bases.

pH Scale The range of units that indicate whether a substance is acidic, basic, or neutral. The pH scale ranges from 0 to 14.

Reactive Having the tendency to chemically combine with something else and change its form. For example, a strong acid is highly reactive with a strong base.

The pH Scale

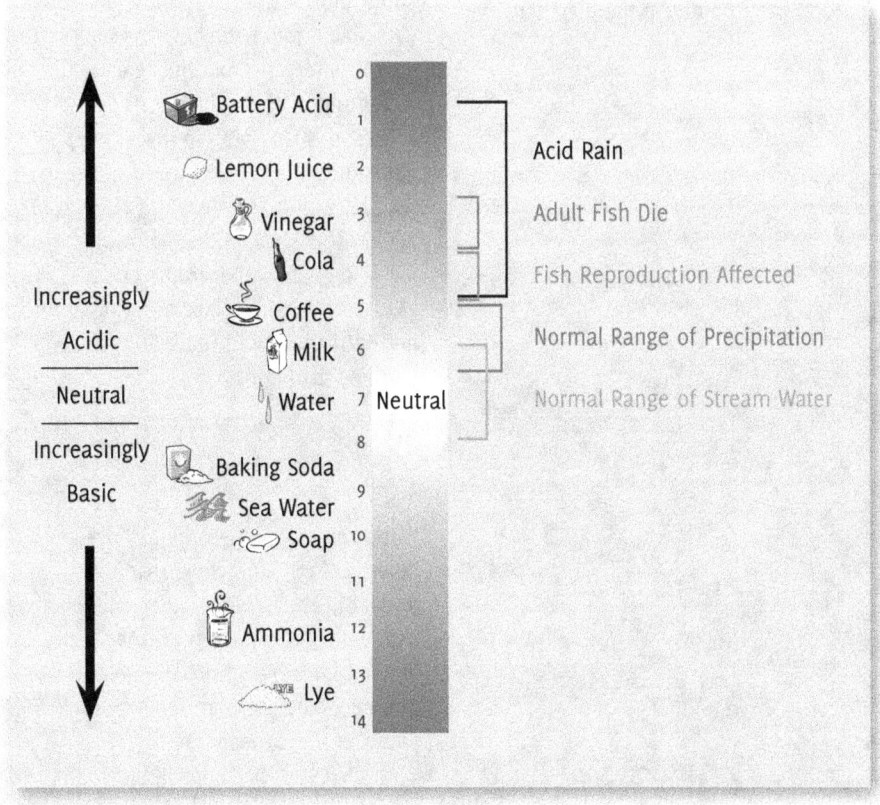

Try These Experiments!

Experiment 1
Measuring pH   p.23

Experiment 2
Determining the pH of
Common Substances   p.24

Experiment 3
Making a Natural
pH Indicator

p.25

Defining Acid Rain

TERMINOLOGY

Atmosphere The air or gases that surround a planetary body such as the Earth.

Sulfuric Acid An acid that can be produced in the atmosphere from sulfur dioxide, a pollutant that results from burning fossil fuels.

Nitric Acid An acid that can be produced from nitrogen oxide, a pollutant that results from the burning of fossil fuels.

Precipitation Water falling to the Earth. Mist, sleet, rain, hail, fog, and snow are the most common kinds of precipitation.

Hydrologic Cycle The movement of water from the atmosphere to the surface of the land, soil, and plants and back again to the atmosphere.

Acid rain is rain that is more acidic than it should be. Acid rain is a complicated problem affecting soil and water chemistry, as well as the life cycles of plants and animals on land and in the water. In addition, weather conditions contribute to air pollution and cause acid rain to spread vast distances.

Air Pollution Causes Acid Rain

Scientists have discovered that air pollution from the burning of fossil fuels is the major cause of acid rain. Power plants and factories burn coal, oil, and natural gas to produce the electricity we need to do all kinds of things, like light our homes. Cars, trucks, and airplanes also run on gasoline, a fossil fuel.

When we burn things, they do not disappear. For example, when you burn a log in a campfire, ash is left. But what happened to the rest of the log? Water from the log becomes vapor and enters the air. Burning wood also releases chemicals and particles into the air. The same thing happens when we burn fossil fuels. Burning fossil fuels sends smoke and fumes into the ATMOSPHERE, or the air above the Earth. In the air, these pollutants combine with moisture to form acid rain. The main chemicals in air pollution that create acid rain are sulfur dioxide (SO_2) and nitrogen oxides (NO_X). Acid rain usually forms high in the clouds where SO_2 and NO_X react with water and oxygen. This forms SULFURIC ACID and NITRIC ACID in the atmosphere. Sunlight increases the speed of these reactions, and therefore the amount of acid in the atmosphere. Rainwater, snow, fog, and other forms of PRECIPITATION then mix with the sulfuric and nitric acids in the air and fall to Earth as acid rain.

Acid Precipitation

Water moves through the air, streams, lakes, oceans, and every living plant and animal in the HYDROLOGIC CYCLE, shown in the image to the left. In that cycle, water EVAPORATES from the land and sea and becomes a gas in the atmosphere. Water in the atmosphere then CONDENSES, or becomes liquid again,

Hydrologic Cycle

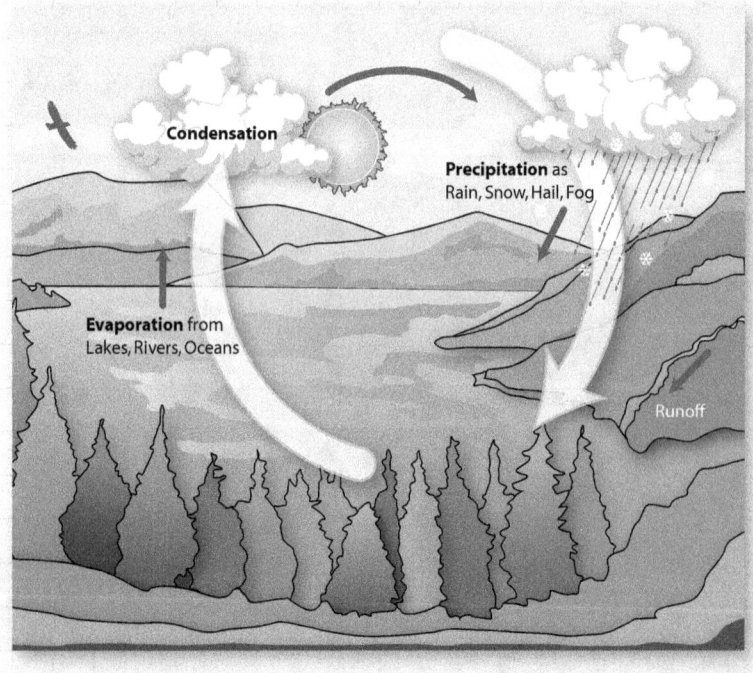

Condensation

Precipitation as Rain, Snow, Hail, Fog

Evaporation from Lakes, Rivers, Oceans

Runoff

Tall smokestacks send pollution high into the air. The longer the pollution is in the air, the greater the chances that the pollutants will form acid rain.

TERMINOLOGY

Evaporate To change from liquid into gas.

Condense To change from gas or vapor to liquid form.

Dry Deposition The falling of small particles and gases to the Earth without rain or snow.

Runoff Water that flows off land into lakes and streams.

Acid Deposition Acidic material that falls from the atmosphere to the Earth in either wet (rain, sleet, snow, fog) or dry (gases, particles) forms.

and forms clouds. Clouds release the water back to the Earth as rain, sleet, hail, snow, or fog. When water droplets form and fall to the Earth they pick up particles like the dust and chemicals that float in the air. Even clean, unpolluted air contains particles such as dust or pollen. Clean air also contains naturally occurring gases such as carbon dioxide (CO_2). The interaction between the water droplets and the CO_2 in the atmosphere gives rain a pH of 5.6, making even clean rain slightly acidic. However, when rain contains pollutants, especially SO_2 and NO_x, the rainwater can become very acidic.

Dry Deposition

Acid rain does not account for all of the acidity that falls back to Earth from pollutants. About half of the acidity in the atmosphere is deposited onto buildings, cars, homes, and trees—anything!—as particles and gases. This process is called DRY DEPOSITION. In some instances, these gases and particles can damage or alter the things on which they settle. Dry deposition (gases and particles) is sometimes washed from trees and other surfaces by rainstorms. When that happens, the RUNOFF water contains acid from acid rain and dry deposition, making the combination more acidic than the falling rain alone. The combination of acid rain (wet deposition) plus dry deposition is called ACID DEPOSITION.

Acid Rain Is A Problem That Can Travel

The chemical reactions that cause acid rain can take several hours to several days to occur. Years ago, when smokestacks were only a few stories high, pollution from smokestacks usually stayed near the ground and settled on the land nearby. This caused unhealthy conditions for people, plants, and animals near those smokestacks. To reduce this pollution, the government

Dry deposition can be washed away from surfaces such as buildings and cars during rainstorms.

Try This Experiment!

Experiment 4
Measuring the pH Of Natural Water
p.26

Volcanoes are a natural source of acid.

passed laws for the construction of very tall smokestacks. At that time, people thought that if the pollution were sent high into the air it would no longer be a problem. Scientists now know that this is incorrect. In fact, sending pollution high into the sky increases the time that the pollution stays in the air. The longer the pollution is in the air, the greater the chances that the pollutants will form acid rain. In addition, the wind can carry these pollutants for hundreds of miles before they become joined with water droplets to form acid rain. For that reason, acid rain, or wet deposition, can be a problem in areas far from sources of pollution. Dry deposition is usually greater near the cities and industrial areas where the pollutants are released.

Natural Acids

There are also natural sources of acids such as volcanoes, geysers, and hot springs. Nature has developed ways of recycling these acids by absorbing and breaking them down. These natural acids contribute to only a small portion of the acidic rainfall in the world today. In small amounts, these acids actually help dissolve nutrients and minerals from the soil so that trees and other plants can use them for food. Unfortunately, the large amounts of acids produced by human activities overload this natural acidity and throw ecosystems off balance.

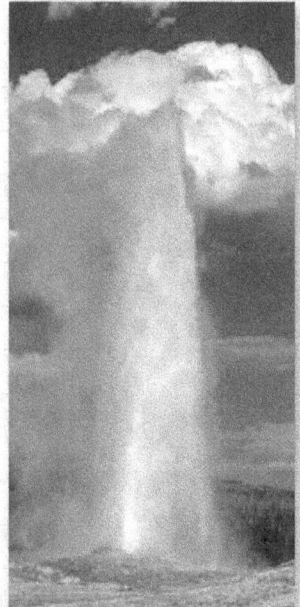

A geyser, like the one shown here in Yellowstone National Park, is also a natural source of acid.

Formation of Acid Rain

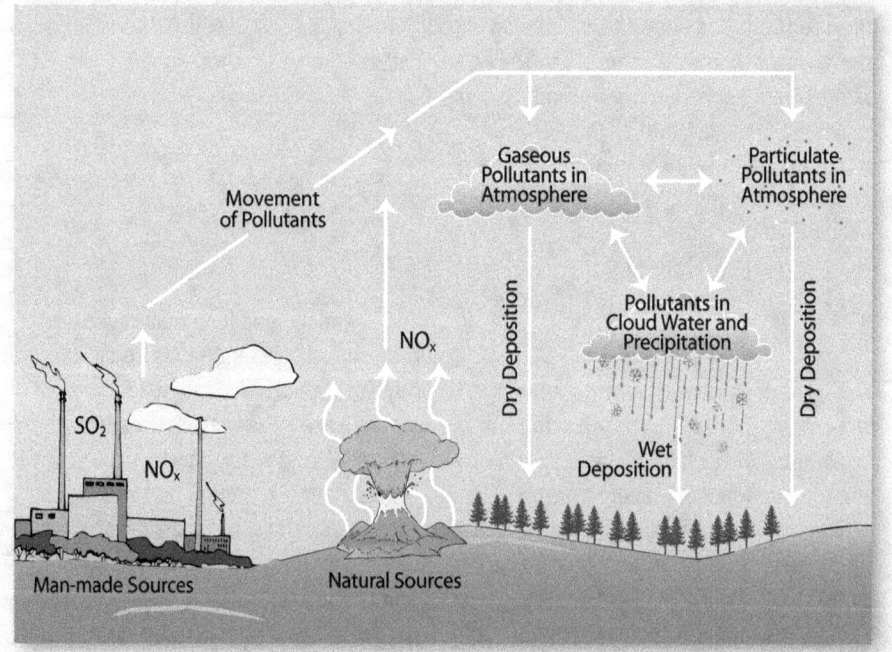

Effects Of Acid Rain On Ecosystems

Acid rain and the air pollution that causes it can severely damage ECOSYSTEMS. An ecosystem is all the living and nonliving things in an area, as well as the interactions between them. Ecosystems come in all sizes. An entire forest is an ecosystem, but so is a single tree. Some scientists even consider the entire Earth an ecosystem. The study of ecosystems is called ECOLOGY. Ecologists study things like predator-prey relationships, how nutrients are taken from the soil into trees, or the kinds of bacteria found in a pond. Every ecosystem is very interconnected, and the organisms that live there rely heavily on each other. For example, ecosystems have food webs, where species depend on one another for food. If any one animal is affected, so are several others. This is how acid rain can affect entire ecosystems. Acid rain may only damage a few organisms in an ecosystem, but everything else is indirectly affected. The damage acid rain causes can also take years, or even decades to reverse.

Forests

Acid rain causes significant damage to forests. It directly affects trees and other plants which are important to the ecosystem as a whole because they are PRIMARY PRODUCERS. Primary producers are organisms that produce their own food through PHOTOSYNTHESIS, a series of chemical reactions that convert water into sugar using light from the sun to provide energy. Plants and some microscopic animals have this ability. Plants

Sugar Maple leaves turn brilliant shades of red, orange, and yellow in the fall. People from all over the United States and the world travel to New England to see colorful autumn leaves like those of the Sugar Maple.

are important to ecosystems because they feed everything else, and provide important HABITAT for other animals. If trees and plants are damaged by acid rain, the effects are felt throughout the entire ecosystem.

Acid rain causes trees in forests to grow more slowly, and in some sensitive species it can even make the leaves or needles turn brown and fall off. Red Spruce and Sugar Maple, two species of trees found mainly in the East and in New England, are very susceptible to acid rain damage. Acid rain damages trees by dissolving the calcium in the soil and in the leaves of trees. This hurts the tree, because calcium is a mineral that trees need to grow. Once the calcium is dissolved, the rain washes it away so the trees and other plants cannot use it to grow. Acid rain washes other minerals and nutrients from the soil in a similar fashion, causing NUTRIENT DEFICIENCY. This is why acid rain can cause trees to grow more slowly.

Aluminum A silvery-white metal; it is the most abundant metal in the Earth's crust.

Haze When particles of dust, pollen, or pollution make the air less clear, and limit visibility.

Neutralize To combine acids and bases to make a neutral substance or solution. For example, acidic water can be neutralized by adding a base.

Buffering Capacity The ability of a substance to resist changes in pH when acids or bases are added.

Buffer A substance, such as soil, bedrock, or water, capable of neutralizing either acids or bases.

Rain falls through the trees to the forest floor and runs into streams, rivers, and lakes.

Nutrient deficiency causes other problems for trees and plants. The lack of nutrients weakens the trees, and makes them more sensitive to the cold. A well-nourished tree in healthy soil will survive even a very cold winter with little difficulty, but a tree already weakened by a mineral deficiency can die during a cold winter. The weakened trees and plants are also more sensitive to insects and disease.

At the same time, acid rain causes the release of substances such as ALUMINUM from the soil. Aluminum can be very harmful to trees and plants. Once released into soil, aluminum can end up in streams, rivers, and lakes, where it can harm or even kill fish. Less aluminum is released when the rainfall is cleaner.

The pollution that causes acid rain also causes HAZE by scattering light back towards the sky. Haze reduces the amount of light available for plants to use in photosynthesis. Since photosynthesis is the base of the food chain, acid rain can cause problems with the movement of nutrients to other organisms in ecosystems that are already impacted.

Further reducing the amount of photosynthesis are acid fogs. Fog can often be more acidic than rainfall. When leaves are frequently bathed in acid fog, their protective waxy coating can wear away. The loss of this coating damages the leaves and creates brown spots. The leaves are then unable to use photosynthesis to turn the energy in sunlight into food for growth. When leaves are damaged, they cannot produce enough food energy for the tree to remain healthy.

Acid Rain On The Forest Floor

A spring shower in the forest washes leaves and the rain falls through the trees to the forest floor below. Some of the water soaks into the soil. Some trickles over the ground and runs into a stream, river, or lake. Soil sometimes contains substances, like limestone, that buffer acids or bases. Some salts in soil may also act as buffers. The soil may NEUTRALIZE, or make less acidic, the acid rainwater. This ability of the soil to resist pH change is called BUFFERING CAPACITY. A BUFFER resists changes in pH. Without buffering capacity, soil pH would change rapidly. Midwestern states like Nebraska and Indiana have soils that are well buffered. Places in the mountainous northeast, like New York's Adirondack Mountains, have soils that are less able to buffer acids. Other soils, like those in

Try These Experiments!

Experiment 5   p.27
Measuring Soil pH

Experiment 6   p.28
Soil Buffering

Experiment 7
Observing the
Influence of Acid
Rain on Plant Growth   p.29

the Southern Appalachian Region, hold acids from acid rain, making them more susceptible to damage from acid rain. Since there are many natural sources of acids in forest soils, soils in forest areas are especially sensitive to effects from acid rain.

Ponds, Lakes, And Streams

The effects of acid rain are most clearly seen in AQUATIC environments such as streams, lakes, and marshes. Acid rain flows to streams, lakes, and marshes after falling on forests, fields, buildings, and roads. Acid rain also falls directly on aquatic habitats.

Most lakes and streams have a pH between 6 and 8, because the buffering capacity of soil usually neutralizes slightly acidic, clean rain. Lakes and streams become acidic (pH value goes down) when the rainwater itself is so acidic that the surrounding soil cannot buffer the rain enough to neutralize it. For this reason, some lakes in areas where soil does not have a lot of buffering capacity are naturally acidic even without acid rain. In areas like the northeastern United States where soil buffering is poor, acid rain has made already slightly acidic lakes very acidic, with some lakes having a pH value of less than 5. As lakes and streams become more acidic, the numbers and types of fish and other aquatic plants and animals that live in these waters decrease. Some types of plants and animals are able to tolerate acidic waters. Others, however, are acid-sensitive and will leave or die as the pH declines. Some acidic lakes have no fish, because at pH 5 most fish eggs cannot hatch. At lower pH levels, adult fish can die. Substances like aluminum that wash into the water from the soil can also harm and kill fish.

Snails have a very low pH tolerance of 6.0.

TERMINOLOGY

Aquatic Relating to water.

Some animals can survive in water that is moderately acidic, while other animals can only live in water that is near neutral. An animal that can survive in moderately acidic water is said to have a high tolerance for acidity. The chart to the left shows the pH tolerance of various animals. Where the boxes are gray, the animals can survive, but where the boxes are black, the animals can no longer tolerate the acidity and die. From the chart, you can see that frogs have a high tolerance for acidity, while clams and snails have a low tolerance.

pH Tolerance Chart on Aquatic Life

	pH 6.5	pH 6.0	pH 5.5	pH 5.0	pH 4.5	pH 4.0
Trout						
Bass						
Perch						
Frogs						
Salamanders						
Clams						
Crayfish						
Snails						
Mayfly						

Effects Of Acid Rain-Causing Pollutants On Humans

These are photos from the Great Smoky Mountains National Park in Tennessee and North Carolina. The photo on the left, taken January 21, 1986, shows what the park looks like on a clear day, or a day where there is little pollution present to reduce visibility. The photo on the right, taken August 5, 1986, shows the same view on a hazy day, when the air is filled with aerosols.

Air pollution, shown here in New York City, can cause health problems.

Acid rain looks, feels, and tastes just like clean rain. Walking in acid rain, or even swimming in an acid lake, is no more dangerous for humans than walking or swimming in clean water. However, breathing air that contains the pollutants that cause acid rain can damage human health. Sulfur dioxide (SO_2), nitrogen oxides (NO_x), particulate matter, and ozone all irritate or even damage our lungs. These effects are mostly seen in people whose lungs have already been weakened by RESPIRATORY ILLNESS, but even healthy people can sometimes have pain or difficulty breathing because of air pollution.

Ozone is a dangerous pollutant that is caused by air pollution, especially in the summer. Exposure to high levels of ozone have been linked to a number of health problems. Ozone can make respiratory illnesses, such as asthma, emphysema, and bronchitis worse. Ozone can also reduce the RESPIRATORY SYSTEM'S ability to fight off bacterial

infections. Even healthy people can have symptoms related to ozone exposure, including coughing, pain with deep breathing, chest tightness, and shortness of breath. Over time, ozone can cause permanent damage to the lungs or even death. Small particles called particulate matter are made up of the same pollutants that cause acid rain. Particulate matter also damages the lungs. The tiny particles of dust that make up particulate matter can bypass the body's natural defenses and become lodged deep in the lungs, where it can cause irritation and damage the lungs.

SO_2 and NO_x, the pollutants that cause acid rain, can also reduce visibility, limiting how far into the distance we can see. These pollutants form small particles in the atmosphere. These particles reduce visibility by scattering light. Reduced visibility is most noticeable in places like National Parks, where people go to see some of the nation's most beautiful landscapes.

Effects Of Acid Rain On Man-Made Materials

Acid rain eats away at stone, metal, paint—almost any material exposed to the weather for a long period of time. Human-made materials gradually deteriorate even when exposed to unpolluted rain, but acid rain speeds up the process. Acid rain can rust metals and cause marble statues carved long ago to lose their features. This happens because marble is made of a compound called calcium carbonate, which can be dissolved by acids. Calcium carbonate is also found in limestone. Many buildings and monuments are made of marble and limestone and are damaged by acid rain. Repairing acid rain damage to buildings and monuments can cost billions of dollars. Historical monuments and buildings, such as the Lincoln Memorial in Washington, D.C., can never be replaced.

How Acid Rain Affects Stonework

The picture on the top was taken in 1908. The picture on the bottom was taken in 1968.

How Acid Rain Affects Metal

The Lincoln Memorial in Washington, D.C.

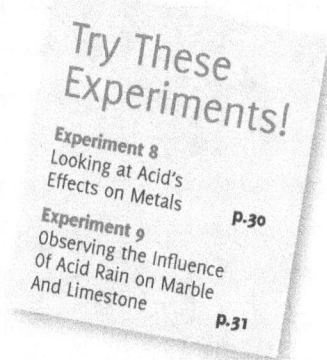

Try These Experiments!

Experiment 8
Looking at Acid's Effects on Metals   p.30

Experiment 9
Observing the Influence Of Acid Rain on Marble And Limestone
                                                       p.31

13

What Is Being Done

The Rockport Power Plant in Rockport, Indiana, burns coal to produce electricity.

EPA has been working with Congress and other federal government agencies; state, local, and tribal governments; scientists; and citizens to solve the acid rain problem for over 15 years.

The Acid Rain Program

The Acid Rain Program was established by Congress as part of the 1990 Clean Air Act Amendments. It requires the electric power industry to lower emissions of sulfur dioxide (SO_2) and nitrogen oxides (NO_x), the pollutants that cause acid rain.

By the year 2010, power plants must reduce emissions of these pollutants by about 50 percent from levels in 1980. Today, power plants emit 35 percent less SO_2 and 46 percent less NO_x than they did in 1990.

The Acid Rain Program is so successful because it uses a CAP AND TRADE program to reduce emissions. A cap and trade program is a policy that controls large amounts of emissions from a group of sources. The approach first sets an overall CAP, or maximum amount of emissions allowed, for all pollution sources under the program. The cap is chosen in order to meet an

Cap and Trade

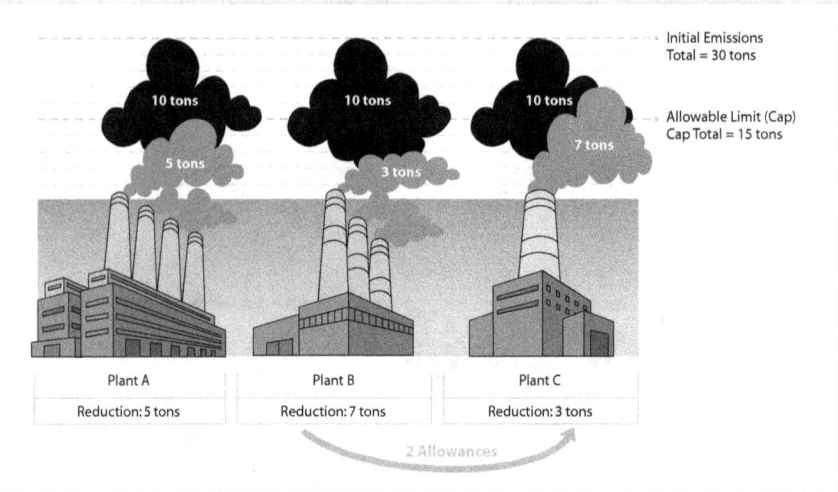

All three plants were required to reduce their emissions from 10 tons to 5 tons each. Plant A reduced its emissions by 5 tons to meet the cap and be in compliance. Plant B reduced its emissions to 3 tons, 2 tons more than were required. Plant C only reduced its emissions to 7 tons. This is not enough to cover its allowances and be in compliance. In a cap and trade program, Plant B can sell its two extra allowances to Plant C. Plant C can then use the two allowances it bought to cover its emissions and stay under its cap.

Notice that even though a trade took place, overall emissions for Plants A, B, and C are within the cap ($5 + 3 + 7 = 15$). Total emissions in this cap and trade system are half of what they used to be (30 tons vs. 15 tons).

environmental goal. In this case the cap is the maximum level of SO_2 and NO_x emissions that all power plants combined can emit. ALLOWANCES, or permits to pollute a specific amount, are then given to power plants. It is up to the individual power plants to decide how they want to lower their emissions, but the total amount of emissions from all power plants across the country must be less than the cap. If one source reduces their emissions far below the amount assigned to them, they are allowed to sell their leftover emission allowances to another power plant that did not reduce their emissions enough. The option of selling extra emission allowances to make money is an incentive for the power plants to reduce their emissions even more than required. Companies that do not reduce their emissions enough and emit more pollution than they have allowances for, must buy allowances to cover their emissions or face heavy fines and other strict penalties. This rarely happens in the Acid Rain Program because the strict penalties discourage power plants from not having enough allowances. In fact, in 2006, the Acid Rain Program had 100 percent compliance from sources of emissions.

Power plants can reduce the amount of pollution they produce in several ways. Some plants choose to wash the sulfur out of coal before it is burned. There are also different kinds of coal and some have less sulfur and nitrogen in them. Pollutants can also be removed from the smoke as it travels through the smokestack. A device called a SCRUBBER removes sulfur from the smoke by spraying a mixture of water and powdered limestone into the smokestack. This mixture traps the sulfur before it can escape into the air above. There are also several other ways to decrease the pollution coming from power plants and scientists and engineers are always discovering new ones.

Wet Acid Deposition, 1989-1991   Wet Acid Deposition, 2003-2005

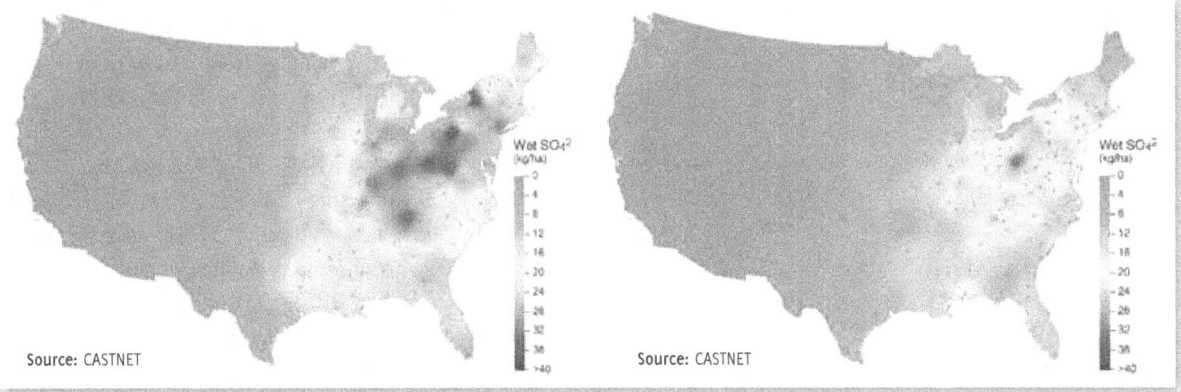

Source: CASTNET   Source: CASTNET

These maps show how the Acid Rain Program has reduced the amount of wet acid deposition (acid rain) in the United States. In these maps, the dark areas represent places with high amounts of acid rain. Notice how the areas with lots of acid rain in 1989–1991 have shrunk since the Acid Rain Program started.

15

Atmospheric Relating to the atmosphere, or the air above the Earth.

Meteorological Pertaining to the weather.

A CASTNET Monitoring Site

Cap and trade is most effective when the environmental or public health issue occurs over a large area, when there are many sources of pollution which contribute to the problem, when emissions can be monitored, and when there is variety in the amount of money that sources must pay to reduce their emissions.

A mandatory cap on emissions is critical to protect public health and the environment. The Acid Rain Program and the NO_x Budget Trading Program have been very effective in reducing emissions of SO_2 and NO_x. Though long-term environmental monitoring has proven that these programs are working, studies have shown that more reductions in emissions are necessary to protect human health and the environment. In 2009, a new program called the Clean Air Interstate Rule (CAIR) will begin. This program will lower power plant

emissions of SO_2 and NO_x in the eastern United States even further than the Acid Rain Program.

To learn more about cap and trade, go to the Clean Air Markets cap and trade Web site at www.epa.gov/airmarkets/cap-trade/. If you want to know more about the other cap and trade programs, like the NO_x Budget Trading program or CAIR, go to the Clean Air Markets program Web site at www.epa.gov/airmarkets.

Monitoring

How does EPA know its programs are working? Experts from EPA, states, universities, and other agencies have set up air quality and deposition monitoring stations across the country. These monitoring stations contain equipment that constantly collects air quality data and samples. These devices measure many things, including the amount of pollution in the air, the pH of rain, the amount of rainfall, and the surrounding temperature. There are several networks made up of many stations taking samples in different areas. The Clean Air Status and Trends Network (CASTNET) takes samples from mostly rural areas around the United States. CASTNET measures dry deposition and collects ATMOSPHERIC data. The National Atmospheric Deposition Program (NADP) has sites around the United States and focuses on precipitation and METEOROLOGICAL monitoring. By using this information, EPA is able to track the success of the Acid Rain Program and other cap and trade programs by linking

CASTNET Monitoring Stations

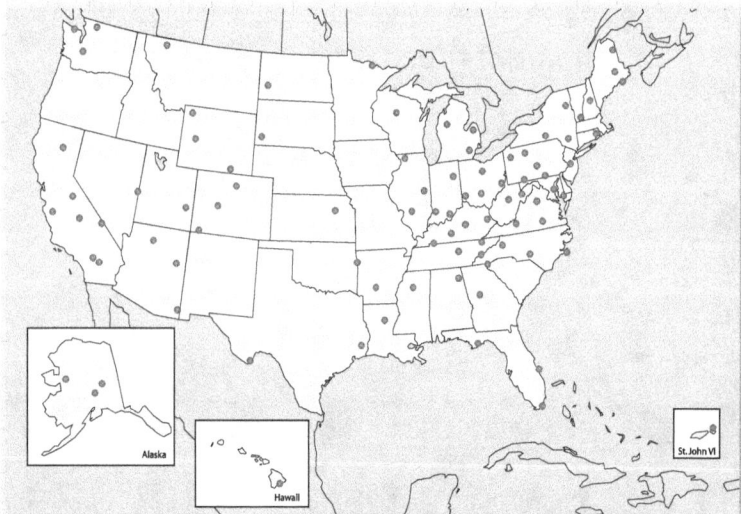

CASTNET now has over 80 monitoring stations across the United States.

reductions in emissions to improvements in air quality. You can see these monitoring stations and the information they collect online at http://camddataandmaps.epa.gov/gdm/.

EPA also requires power plants to use Continuous Emissions Monitoring Systems (CEMS) to keep track of the amount of pollution they release into the air. A CEMS is a monitoring device that each unit must place on their smokestack. These monitors take samples of the air traveling through the smokestack, and measure the amount of pollutants traveling through it. Then, the monitor sends the recordings to EPA. CEMS allows EPA to keep track of emissions to make sure that the power companies are following the laws to reduce pollution.

Wind Power: Windmills produce electricity.

Alternative Ways Of Producing Energy

There are other sources of energy besides fossil fuels. These include HYDROELECTRIC POWER, WIND POWER, NUCLEAR POWER, SOLAR POWER, and FUEL CELLS. Hydroelectric dams use the power of water to turn TURBINES and make electricity. Windmills work the same way but instead, use wind to turn the turbines. People have been using wind and water power for centuries. Nuclear power plants collect the energy released by splitting tiny atoms apart inside nuclear reactors. Although nuclear power plants generate dangerous waste that must be disposed of carefully, a small amount of nuclear fuel can make a very large amount of electricity. Some people also use solar power, or power from the sun, to make electricity. Some houses use solar power to heat water for showers, and even some traffic signs run off of solar panels. Fuel cells are similar to batteries, except that fuel cells run on oxygen and hydrogen. They use chemical reactions to generate electricity, and produce water as a waste.

All sources of energy have benefits and limitations, including the cost of producing the energy. All of these factors must be weighed when deciding which energy source to use.

Solar Power: Solar panels produce electricity.

TERMINOLOGY

Hydroelectric Power Energy that is generated by dams, which use water to turn turbines and generate electricity.

Wind Power Energy that is generated when the wind turns the sails of a windmill, which are attached to turbines that generate electricity.

Nuclear Power Energy that comes from breaking apart the center (nucleus) of an atom.

Solar Power Electricity that is generated by harnessing the energy of the sun. Solar panels are often used to convert sunlight into energy.

Fuel Cells Similar to batteries, fuel cells store energy that can be used to power all sorts of things. Unlike a battery though, fuel cells do not "run out" and do not need to be recharged or replaced.

Turbine A motor activated by water, steam, or air to produce energy.

What You Can Do To Help

All environmental problems, including acid rain, are caused or impacted by the combined actions of individual people. This is why individuals can play a big part in solving these problems. While you cannot run a nationwide cap and trade program from your classroom, there are many things you can do to help reduce pollution and protect the environment.

To Prevent Acid Rain

1. Conserve electricity by turning off lights, appliances, and computers when no one is using them.

2. Join ENERGY STAR®'s "Change a Light, Change the World" program and pledge to use energy-efficient light bulbs in your home. This will reduce the amount of energy you use, and in turn reduce the amount of emissions produced by power plants. You can also lower your energy consumption by using other Energy Star® products and appliances, including TVs, computers, refrigerators, washing machines, air conditioners, furnaces, etc. For more information, see EPA's Energy Star® Web site www.energystar.gov.

3. You can greatly reduce acid rain-causing emissions (especially NO_x emissions) by changing your transportation habits. Try to reduce the amount of time you spend in the car by walking, biking, or taking public transportation. When you drive, plan trips ahead of time to minimize miles traveled, drive the most fuel-efficient car you can, and carpool whenever possible. The EPA publishes a green vehicle guide at www.epa.gov /greenvehicles/.

4. Shrink your "carbon footprint" and reduce greenhouse gas emissions by turning the thermostat down a little bit in winter, or up a little bit in the summer. New technologies can also greatly increase the efficiency of your heating and air conditioning and other appliances. Look for the ENERGY STAR® Label. EPA provides information about some of these technologies and other ways to reduce greenhouse gas emissions at www.epa.gov/climatechange/wycd.

Taking public transportation can help reduce acid rain-causing emissions.

To Address Other Environmental Problems

1. Volunteer to help clean up trash and litter in local streams and rivers, or help restore wetlands. By cleaning up trash, you help protect ecosystems that are important habitats to wildlife. Information about wetland restoration efforts in your area can be found in the restoration project directory at www.epa.gov/owow /wetlands/restore/.

2. Recycle everything you can at home and if your school does not already have a recycling program, start one! This reduces the amount of waste that you produce, and keeps recyclable materials out of landfills. Also, try to buy products with less packaging, or products with recyclable packaging. More information about recycling and waste reduction programs can be found at EPA's solid waste education Web site, www.epa.gov /epaoswer/education.

3. Many areas of the United States, especially the Southwest, have problems with water availability. Conserving water helps ecosystems and the people and animals that live in them. You can help by reducing the amount of water you use by taking shorter showers, or turning off the faucet while brushing your teeth. Learn more by visiting the EPA's Water Sense page at www.epa.gov /owm/water-efficiency.

Recycling at home is a great way to keep recyclable materials out of landfills.

4. Try talking to your school's principal or superintendent about ways to reduce the impacts your school has on the environment. There are many small things your school can do to help protect ecosystems. For some ideas, check out http://cfpub.epa.gov/schools.

5. Spread the word! Share what you have learned about Acid Rain and other environmental problems with others. Tell them what they can do to protect the environment. The more information people have, the more they can do to make the Earth a cleaner, healthier place.

Experiments

pH Paper

TERMINOLOGY

Litmus Paper Paper coated with a chemical coloring obtained from lichens that turns red in acidic water and blue in basic water. It is used as an acid-base indicator.

pH Paper Paper that changes color to show the pH of a substance.

The answers to all of the questions in the experiments section can be found at the end of the section, on page 33.

Measuring With pH Paper

For most of the following experiments, you will need a pH indicator. A pH indicator contains a chemical that changes color when it is exposed to acids or bases. For example, LITMUS and pH PAPER turn red in strong acids and blue in strong bases. Because only a few pH indicators measure pH over a wide range of pH values, you will need to find out the pH range of the indicator you use. Typically, the color chart provided with each pH indicator kit will show the pH range of that indicator. Color pH indicators provide only an approximate measure of the pH, or the strength of the acid or base. They are not as accurate as pH meters, but they are adequate for the following experiments.

When measuring pH with pH paper or litmus paper, dip the end of a strip of the paper into each mixture you want to test. Follow the directions on the package regarding how long you need to keep the pH paper in the mixture and how long to wait before reading the measurement. Then compare the color at the wet end of the paper with the color chart provided with that pH indicator. Write down the pH value and color. Always use a clean, unused strip of pH paper for each mixture that you test. Be sure to conduct the pH test for each substance three times, using a new pH paper for each test. Record the results of each test. If possible, have a different student conduct and record each test. This helps to ensure scientific accuracy, consistency, and replicability.

Measuring With pH Meters

A pH meter provides a more precise pH measurement than pH paper. Before using a pH meter, rinse the electrode with distilled water and blot dry with a clean paper towel. Calibrate the meter according to the directions. When testing the pH of a substance, put the electrode tip in the mixture and stir once. Be sure not to touch the bottom or sides of the container. Hold the electrode in the mixture for 1 minute or until the reading is steady. Record the measurement and repeat the test two more times for accuracy and consistency.

For more information on measuring the pH of soil and water, check out the pH protocols defined by GLOBE (Global Learning and Observations to Benefit the Environment) at www.globe.gov/protocols.

Tips

- Except for wide-range pH test paper and pH meters, all the materials called for in these experiments can be obtained at grocery stores or from local lawn and garden stores or nurseries.

- Wide-range pH test paper is pH paper that covers the whole pH scale. Not all pH test papers do this. Other papers cover only part of the pH scale and there are different papers to test for acids, bases, and neutrals. Both kinds are inexpensive, and a school science laboratory will probably have one or the other, if not both. If the school does not already have pH test paper a science teacher may know where to order it, or you may order it on your own through a biological supply company. If you have to order pH test paper, we recommend wide range pH test paper since it can be used for all the experiments, and may be less confusing to younger students.

- Baking soda and ammonia are both bases. You may substitute baking soda for household ammonia in the experiments. If you do, be sure to stir well because baking soda does not dissolve easily in water unless heated. The pH of undissolved baking soda will not be the same as dissolved baking soda.

- Lemon juice and white vinegar are both acids. You may substitute fresh-squeezed lemon juice for white vinegar. Lemon juice is slightly more acidic than the vinegar sold in grocery stores. White vinegar is preferred over cider vinegar or lemon juice because it is colorless and relatively free of impurities.

- Use clean, dry containers and utensils.

Safety In The Laboratory

A science or chemistry laboratory can and should be a safe place to perform experiments. Accidents can be prevented if you think about what you are doing at all times, use good judgment, observe safety rules, and follow directions.

Always wear protective safety glasses or goggles when working on experiments.

Writing down your observations can be very helpful.

- Eye protection (goggles or safety glasses) must be worn when working on experiments. Make a habit of putting them on before the experiment begins and keeping them on until all clean up is finished.

- Do not eat or drink while in the laboratory.

- Do not taste any chemical.

- Long-sleeved shirts and closed-toe, leather-topped shoes must be worn at all times.

- Long hair must be tied back, so it will not fall into chemicals or flames.

- Do not work alone; work with an adult.

- Never perform any unauthorized experiment.

- All glassware must be washed and cleaned. Wipe all counter surfaces and hands with soap and water.

- All experiments that produce or use chemicals must be done in a well-ventilated area.

- Never point the open end of a test tube at yourself or another person.

- If you want to smell a substance, do not hold it directly to your nose. Instead, hold the container a few centimeters away and use your hand to fan vapors toward you.

- When diluting acids, always add the acid to the water; never water to acid. Add the acid slowly.

- Dispose of all chemicals properly, according to the directions of your teacher.

- If you spill any acid or base material on you, wash the exposed area with large amounts of cold water. If skin becomes irritated, see a physician.

Recording Observations

Writing your observations on these experiments will help you to keep better track of the progress of the experiment and help you to remember information for answering questions or writing lab reports. Record keeping can be very simple and still be a help. These hints can help you organize and record your thoughts.

- Use a bound notebook so that pages are not lost.

- Write complete sentences for all written entries.

- Use drawings as needed.

- Date each entry (even drawings).

- Use the title of the experiment as your first entry.

- When your observation entries have been completed, write your answers to the questions that follow each experiment.

- Write your own thoughts about the experiment as the conclusion.

Experiment ❶
Measuring pH

Materials

- ❏ pH paper and color chart or pH meter (pH range 3 to 12)
- ❏ 1 ½ cups distilled water
- ❏ 1/2 teaspoon white vinegar
- ❏ 1/2 teaspoon household ammonia
- ❏ 3 small cups or beakers
- ❏ 3 clean stirring spoons
- ❏ measuring cups and spoons (1/2 cup and 1/2 teaspoon)
- ❏ notebook and pencil

Instructions

1. Label the first cup "vinegar," the second cup "ammonia," and the third cup "water."

2. Pour 1/2 cup distilled water into each of the 3 cups.

3. Add 1/2 teaspoon white vinegar to the vinegar cup and stir with a clean spoon.

4. Add 1/2 teaspoon ammonia to the ammonia cup and stir with a clean spoon.

5. Do not add anything to the water cup.

6. ✍ Which substance do you think will be an acid? Which one will be a base? Take a moment to write down your hypotheses.

7. Follow the instructions that come with your pH paper or pH meter for testing the pH of the vinegar mixture. Record the pH value.

8. Follow the instructions that come with your pH paper or pH meter for testing the pH of the ammonia mixture. Record the pH value.

9. Follow the instructions that come with your pH paper or pH meter for testing the pH of the distilled water. Record the pH value.

10. Repeat steps 7-9 two more times so you test and record the pH results a total of three times for each mixture.

Questions

1. Is vinegar an acid or a base?

2. Is ammonia an acid or a base?

3. What is the pH of distilled water? Did the pH level surprise you? Why do you think distilled water did not have a neutral pH?

4. Were your hypotheses correct?

This experiment will illustrate how to measure the approximate pH of chemicals in water using a pH indicator.

This experiment will take approximately 20 minutes

Experiment ②

Determining The pH Of Common Substances

In this experiment, you will use a pH indicator to measure the pH of some fruits, common beverages, and soaps. Many foods and household cleaners are either acids or bases.

Materials

- ❏ pH paper and color chart or pH meter (range pH 2 to 12)
- ❏ 3 different fresh, whole fruits (lemon, lime, orange, or melon)
- ❏ 3 different beverages (cola, carbonated non-cola, milk)
- ❏ 1/8 teaspoon dish soap or laundry detergent
- ❏ 1/4 cup distilled water
- ❏ measuring cups and spoons (1/2 cup, 1/4 and 1/8 teaspoon)
- ❏ 4 small cups or beakers
- ❏ 1 clean stirring spoon
- ❏ notebook and pencil
- ❏ paring knife

Instructions

❶ Using the knife, carefully cut each fruit in half, cleaning off the knife after each cut.

❷ Label the 3 cups "cola," "non-cola," and "milk."

❸ Pour about 1/2 cup of each liquid into the appropriately labeled cup.

❹ In the fourth cup add 1/8 teaspoon soap to 1/4 cup distilled water and stir for about 2 minutes.

❺ ✑ Given what you have learned, which items do you think will be acids? Which will be bases? Why? Take a moment to write down your hypotheses.

This experiment will take approximately 40 minutes

❻ Follow the instructions that come with your pH paper or pH meter for testing the pH of each fruit. Record the pH value. Be sure to use a clean, unused pH paper for each one or clean the pH meter electrode between testing each substance.

❼ Follow the instructions that come with your pH paper or pH meter for testing the pH of the cola and other beverages. Record the pH value. Be sure to use a clean, unused strip of pH paper for each one or clean the pH meter electrode between testing each substance.

❽ Follow the instructions that come with your pH paper or pH meter for testing the pH of the soap mixture. Record the pH value.

❾ Be sure to conduct the pH test three times for each fruit or substance and record the results of each test.

Questions

1. Are lemons, limes, oranges, and melons acids or bases?

2. Are colas and non-colas acids or bases?

3. Was the milk acidic or basic?

4. Was the soap/detergent mixture acidic or basic?

5. Which item is most acidic? Which one is most basic? How can you tell?

6. What other foods or drinks are acids? Why?

7. Were your hypotheses correct?

Experiment ❸
Making A Natural pH Indicator

Materials

- ❏ 1 head of red cabbage, sliced
- ❏ stainless steel or enamel pan or microwaveable casserole dish
- ❏ 1 quart water
- ❏ stove, microwave, or hotplate
- ❏ 1/2 teaspoon white vinegar
- ❏ 1/2 teaspoon ammonia
- ❏ 1 teaspoon clear, carbonated beverage, like seltzer water or lemon-lime drinks
- ❏ 3 small clear cups or beakers
- ❏ 3 clean stirring spoons
- ❏ measuring cups and spoons (1 quart, 1/4 cup, 1/2 and 1 teaspoon)
- ❏ notebook and pencil

Instructions

❶ Boil sliced cabbage in 1 quart of water in a covered pan for 30 minutes or microwave for 10 minutes. (Don't let the water boil away.)

❷ Let cool before removing the cabbage.

❸ Label 3 cups "ammonia," "vinegar," and "beverage."

❹ Pour about 1/4 cup of cabbage juice into each cup.

❺ ✍ What color do you think the cabbage juice will turn when you add ammonia? What about when you add vinegar or the beverage? Take a moment to write down your hypotheses.

❻ Add 1/2 teaspoon ammonia to the cup labeled "ammonia" and stir with a clean spoon.

❼ Add 1/2 teaspoon vinegar to the cup labeled "vinegar" and stir with a clean spoon.

❽ Add 1 teaspoon clear non-cola to the last cup and stir with a clean spoon.

❾ Record and observe what happens to the color of the liquid in each cup.

❿ After answering the first two questions for this experiment, pour the contents of the vinegar cup into the ammonia cup.

Questions

1. What color change took place when you added vinegar to the cabbage juice? Why?

2. Did the ammonia turn the cabbage juice pH indicator red or blue? Why?

3. If you were to gradually add vinegar to the cup containing the ammonia and cabbage juice, what do you think would happen to the color of the indicator? Try it, stirring constantly.

4. Is the non-cola soft drink acidic or basic?

5. Were your hypotheses correct?

In this experiment, you will make your own pH indicator from red cabbage. Red cabbage contains a chemical that turns from its natural deep purple color to red in acids and blue in bases. Litmus paper, another natural pH indicator, also turns red in acids and blue in bases.

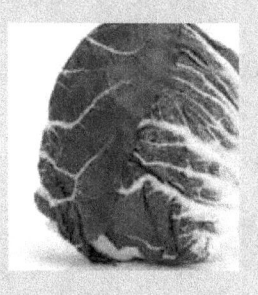

This experiment will take approximately 50 minutes

Experiment ④
Measuring The pH Of Natural Water

TERMINOLOGY

Surface Water Water that is found on the surface of the Earth in lakes, ponds, streams, or rivers.

This experiment will take approximately **15 minutes** in the classroom with additional time for water collection

Materials

- ❏ pH paper and color chart or pH meter (range pH 2 to 7)

- ❏ clean cups or beakers

- ❏ 2 clean, empty, water or soda bottles

- ❏ notebook and pencil

- ❏ A map of the pH of rainwater in the United States, which can be found at http://nadp.sws.uiuc.edu/isopleths. After clicking on annual maps and selecting the most recent year available, select the report titled "Lab pH."

Instructions

① Label a clean, empty water or soda bottle "rainwater," and leave it outside of your classroom to collect water on a rainy day.

② After you have collected at least a tablespoon of water, cap the bottle to prevent evaporation and bring it inside.

③ Locate a local stream, river, lake, or pond. Go with an adult. Depending on where you live, you may choose to gather water samples from many sources, including different creeks, lakes, ditches, ponds, and rivers.

④ Scoop some of the SURFACE WATER into a clean, empty water or soda bottle. Label the bottle "surface water," and cap the bottle to prevent evaporation and spilling while you bring it home.

⑤ ✍ Do you think the rainwater you collected will be acidic? What about the different surface water samples? Why? Take a moment to write down your hypotheses.

⑥ Pour the rain water into a cup and measure the pH of the rainwater using pH paper or a pH meter and record the result.

⑦ Pour the surface water into a cup and measure the pH of the surface water using pH paper or a pH meter and record the result.

⑧ Repeat steps 6 and 7 two more times so you test and record the pH results a total of three times for each sample.

Questions

1. What is the pH of the rainwater? Compare your results to the map. Is it what you expected?

2. What is the pH of the surface water?

3. How does the measured pH compare to the pH levels that affect plants and animals in aquatic habitats? (See chart on page 11.)

4. Was one of your samples more acidic than the other? Which one? If they are different, what do you think the reason for the difference is?

5. Were your hypotheses correct?

Experiment ⑤
Measuring Soil pH

Materials

- ❏ garden soil pH test kit

- ❏ 2 cups soil from each of two or three different locations (some of the soil will be needed for the "Soil Buffering" experiment)

- ❏ measuring spoons

- ❏ digging tool

- ❏ self-sealing plastic bags

- ❏ notebook and pencil

Instructions

❶ Pick two or three different soil locations, such as a garden, the school grounds, a wooded area, city park, meadow, etc. Go with an adult.

❷ At each location, quietly observe the plants and animals living in or rooted on these soils, especially those that are in greatest numbers. Write down your observations.

❸ Dig down about 2 inches, scoop out 2 cups of soil, and seal it in a plastic bag for later use. Be sure to clean your digging tool after collecting soil samples at each location.

❹ Label each plastic bag.

❺ ✎ What do you think the pH level of the soil from each location will be? Why? Take a moment to write down your hypotheses.

❻ Back in the classroom, measure the pH of each soil sample following the directions provided in the garden soil pH test kit, and record the approximate pH of each soil sample. Save the excess soil from each site for use in the "Soil Buffering" experiment.

Questions

1. What were the differences between the plant and animal life at each location?

2. What was the range of pH levels at various sites?

3. Were any of your soil samples acidic?

4. Were any of your soil samples basic?

5. Name some possible reasons for the different pH levels at the sites.

6. How do your plant and animal observations at each location and the pH level of the soil compare to the chart on page 11?

7. Were your hypotheses correct?

In this experiment, you will collect soil and measure its pH. Soil pH is one of several important conditions that affect the health of plants and animals. In addition, you will also be asked to survey the plants and animals that live in the area where you collected the soil. Area surveys provide information about how well plants and animals can live under different conditions.

For this experiment, you will need an inexpensive garden soil pH test kit, which may be obtained from lawn and garden stores or nurseries.

This experiment will take approximately **20 minutes** in the classroom with additional time for soil sample collection

Experiment ⑥
Soil Buffering

In this experiment, you will find out if soil from your lawn, garden, or school can buffer acids. You will observe the pH change of an acid mixture poured over soil in a filter. If the water collected from the filter is less acidic than the original mixture, then the soil is buffering some of the acid. If it does not change, then the soil may not be capable of buffering acids. Since the buffering capability of soils differs, you may want to do this experiment with several different soil types including those collected for Experiment 5, "Measuring Soil pH".

This experiment will take approximately **30-60 minutes** depending on the number of soil samples used

Materials

- ❏ pH paper and color chart or pH meter (pH range 2 to 10)
- ❏ about 2 cups of soil from a garden, wooded area, lawn, or school yard
- ❏ distilled water
- ❏ white vinegar
- ❏ ammonia
- ❏ measuring cups and spoons (1 cup and 1 teaspoon)
- ❏ clean stirring spoon
- ❏ large funnel
- ❏ coffee filters
- ❏ clean cups or beakers
- ❏ notebook and pencil

Instructions

❶ Pour 1 teaspoon of vinegar into 1 cup of distilled water. Stir well.

❷ Check the pH with pH paper or a pH meter. The pH of the vinegar/water mixture should be about 4. If it is below that, add ammonia, stir well, and recheck the pH. If it is above pH 4, add a drop or two of vinegar and again recheck the pH.

❸ Put 1 coffee filter into the funnel, and fill the filter with soil from one location. The soil should be slightly packed down, so that the water runs through it slowly. Dense sandy or clay soils might not need to be packed down, but this may be necessary for looser, loamy soils.

❹ Hold the funnel over a cup and slowly pour the vinegar/water mixture over the soil until some water collects in the cup. The filter may clog quickly, but you need only a small amount of water in the cup to test the pH.

❺ ✎ What do you think the pH of the water will be after it filters through the soil? Why? Do you think it will be different for each of the soil samples? Take a moment to write down your hypotheses.

❻ Check the pH of the collected water using pH paper or a pH meter and record the results.

❼ Repeat step 4 with the other soil samples, washing the funnel and using a new coffee filter for each sample.

❽ Repeat step 6 two more times so you test and record the pH results a total of three times for each soil sample.

Questions

1. How did the pH of the collected water change?

2. Why is it important that soil buffer acid?

3. What does acid rain do to soil, and why is this a problem?

4. Were your hypotheses correct?

Experiment ⑦

Observing The Influence Of Acid Rain On Plant Growth

Materials

- ❏ pH paper and color chart or pH meter (pH range 2 to 8)

- ❏ 3 of the same type of healthy, potted plant

- ❏ 3 one-gallon containers with lids (such as a milk jug), one per group of three to four students

- ❏ 1 pint white vinegar

- ❏ tap water with a pH of 7

- ❏ measuring cup (1 pint)

- ❏ notebook and pencil

- ❏ poster paper, crayons, markers for class presentations

Instructions

❶ Divide the class into three teams.

❷ Give each group a 1-gallon container (such as a milk container).

❸ One team will fill their container with 1 gallon (3.8 liters) of tap water. Label the container "tap water."

❹ Another team will fill their container with 1 pint (0.5 liters) of vinegar and 7 pints (3.3 liters) of tap water. Label the container "slightly acidic."

❺ The third team will fill their container with 2 pints (0.9 liters) of vinegar and 6 pints (2.8 liters) of tap water. Label the container "very acidic."

❻ Test the pH of each mixture ("tap water," "slightly acidic," and "very acidic") three times and record each result.

❼ Each team will receive a potted plant and label their plant the same as their container.

❽ Each team will be responsible for watering their plant from the container with the matching label. Students should water the plants when they need it (every 2-4 days). Make sure each group's plant gets the same amount of water in each watering cycle. A good way to do this is to give each group a cup to use when watering.

❾ Place all three plants in the same spot so that they get the same amount of light.

❿ ✐ What do you think will happen to each plant? Will they differ after a few weeks? In what way? Why? Take a moment to write down your hypotheses.

⓫ Have team members examine their plants every day and write their observations. What do the plants look like? What color are they? Are the leaves dropping? Do they look healthy?

⓬ Continue this activity for 2-3 weeks. Then have students examine the plants and discuss or give presentations on the results of the experiment.

Questions

1. Do the plants watered with acid solutions differ in color or size from the others?

2. How long did it take to see the effects of each mixture?

3. Which plant showed the most effects? Why?

4. Were your hypotheses correct?

Acid rain most often damages plants by washing away nutrients and by poisoning the plants with toxic metals, but it can have direct effects on plants as well. In this experiment, you will observe one of the direct effects of acid rainwater on plant growth. The experiment will take 2-3 weeks.

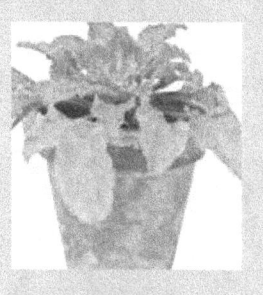

This experiment will take approximately 20 minutes on day 1, 15 minutes a day to record observations, 60-80 minutes for presentations

Experiment ⑧
Looking At Acid's Effects On Metals

When acids and metals are exposed to each other, the metal is gradually dissolved away in a chemical reaction. In this experiment, you will observe this reaction for yourself, but you will need patience. The chemical effect of acids on metals may take at least 5 days for the human eye to see, even though the reaction starts as soon as the acid contacts the metal.

This experiment will take approximately **20 minutes** the first day and **5-10 minutes** each additional day

Materials

- ❏ pH paper and color chart or pH meter (pH range 2 to 7)
- ❏ 2 cups or beakers
- ❏ 2 clean copper pennies (use pennies minted before 1983)
- ❏ 3 tablespoons white vinegar
- ❏ 3 tablespoons distilled water
- ❏ ammonia
- ❏ measuring spoon (1 tablespoon)
- ❏ plastic wrap
- ❏ notebook and pencil

Instructions

① Label one cup "water" and the other "vinegar".

② Put 3 tablespoons of vinegar into the cup, enough to cover the penny completely.

③ Test the pH of the vinegar and record the result.

④ Put 3 tablespoons of distilled water into the cup, enough to cover the penny completely.

⑤ Test the pH of the distilled water. If the pH is below 6, add ammonia and recheck the pH. Repeat this process until the pH is between 6 and 7. Record the pH of the water.

⑥ Repeat steps 3 and 5 two more times so you test and record the pH results a total of three times for each sample.

⑦ Place one penny in each cup. Be sure to use pennies minted before 1983 because pennies minted after that time have a different chemical composition.

⑧ Seal the top of each cup with plastic wrap to prevent evaporation.

⑨ Place in a safe, dry place out of direct sunlight for about 5 days.

⑩ ✍ What do you think will happen to the penny in each cup? How will they differ? Take a moment to write down your hypotheses.

⑪ After 5 days, observe the changes that occurred in each cup and to the penny in each cup.

Questions

1. What change, if any, took place in the distilled water cup and to the penny after 5 days?

2. What change, if any, took place in the vinegar cup and to the penny after 5 days?

3. When you rinsed off the pennies, were you surprised that they both looked about the same as they did at the beginning of the experiment?

4. Were your hypotheses correct?

Experiment ⑨

Observing The Influence Of Acid Rain On Marble And Limestone

Materials

- ❏ pH paper and color chart or pH meter (pH range 2 to 7)
- ❏ 2 pieces of chalk
- ❏ 2 cups or beakers
- ❏ plastic wrap
- ❏ 1/4 cup tap water
- ❏ 1/4 cup white vinegar
- ❏ measuring cup (1/4 cup)
- ❏ notebook and pencil

Instructions

① Label one cup "vinegar" and one cup "water."

② Pour 1/4 cup vinegar into the cup labeled "vinegar."

③ Test the pH of the vinegar and record it.

④ Pour 1/4 cup tap water into the cup labeled "water."

⑤ Test the pH of the water and record it.

⑥ Repeat steps 3 and 5 two more times so you test and record the pH results a total of three times for each mixture.

⑦ Place one piece of chalk in each cup. If you need to add a little more water or vinegar to submerge the chalk, do so.

⑧ Cover each cup with plastic wrap to prevent evaporation, and put in a safe place overnight.

⑨ ✍ What do you think will happen to the piece of chalk in each mixture? Why? Take a moment to write down your hypotheses.

⑩ The next day, observe the two pieces of chalk and record the changes that occurred.

Questions

1. Which piece of chalk is more worn away? Why?

2. Have you seen any buildings in your town that show evidence of acid rain damage? What about the grave markers at an old cemetery? Would you consider acid rain a major problem in your town? Why or why not?

3. Have you seen acid rain damage on any buildings or monuments in other parts of the United States or other countries? Describe what you saw and where it was. If you do not know of any buildings or monuments that have been harmed by acid rain, go online and try to find some pictures of some.

4. Were your hypotheses correct?

Acids do not just eat away at metals. They also eat away at rocks like marble, limestone, and chalk that are made of calcium carbonate. Calcium carbonate is also found in seashells, bones, and teeth. Acids dissolve hard rocks, like marble, very slowly. It can take decades for acid rain damage to become evident on marble structures. Soft rocks, like chalk, can be dissolved much quicker, which is why we use it in this experiment. In this experiment, you will observe how acid can dissolve calcium carbonate.

This experiment will take approximately **15 minutes** each day

Experiment Answers

Below are the answers to the questions provided at the end of each experiment.

Experiment 1: Measuring pH

1. Vinegar is an acid, and in this experiment, it will display a pH of about 4. Vinegar at pH 4 turns pH paper yellow and most other pH indicators red.

2. Ammonia is a base and in this experiment, it will display a pH of about 12. Bases turn most pH indicators blue.

3. Pure distilled water would have tested neutral, but pure distilled water is not easily obtained because carbon dioxide in the air around us mixes, or dissolves, in the water, making it somewhat acidic. The pH of distilled water is between 5.6 and 7. You can neutralize distilled water by adding about 1/8 teaspoon baking soda, or a drop of ammonia, and stirring well. Check the pH of the water with a pH indicator. If the water is still acidic, repeat the process until pH 7 is reached. Should you accidentally add too much baking soda or ammonia, start over or add a drop or two of vinegar, stir, and recheck the pH.

4. Answers will vary.

Experiment 2: Determining The pH Of Common Substances

1. These fruits all contain acids and taste sour. Lemons and limes have pH values near 2. Oranges may be slightly less acidic than lemons and limes, but your pH indicator may not be accurate enough to show the difference.

2. They are both acidic, primarily because they both contain carbon dioxide to make them fizz, and carbon dioxide and water produce carbonic acid. The pH of these beverages varies with the amount of carbon dioxide and other ingredients in them, but it is usually below 4.

3. Milk can be slightly basic or slightly acidic depending on its age and how it was processed at the dairy.

4. Soap contains a base and will turn most pH indicators blue. Alkaline solutions are excellent cleaning agents, which is why we use them to wash dishes and clothes.

5. Answers will vary. You can tell by checking their pH.

6. Answers will vary.

7. Answers will vary.

Experiment 3: Making A Natural pH Indicator

1. The vinegar and cabbage juice mixture should change from deep purple to red, indicating that vinegar is an acid.

2. The ammonia and cabbage juice mixture should change from deep purple to blue, because ammonia is a base, which reacts chemically with the pH indicator, turning it blue.

3. You should find that the acid and base are neutralized, changing the color from blue or red to purple, which is the original, neutral color of the cabbage juice.

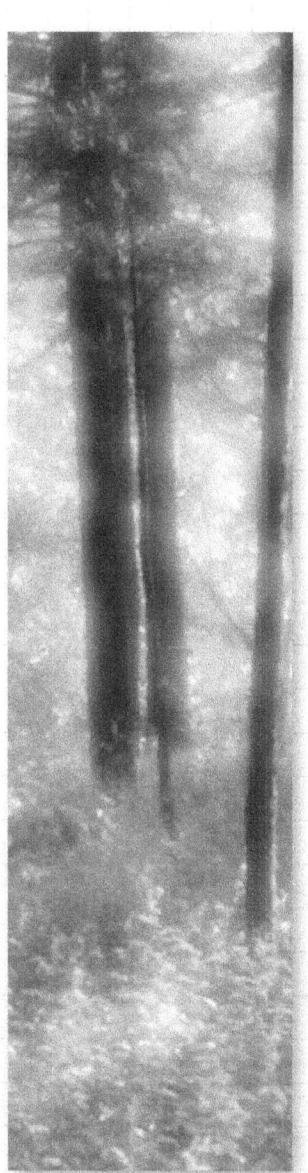

4. The non-cola soft drink is acidic and turns the cabbage juice pH indicator red.

5. Answers will vary.

Experiment 4: Measuring The pH Of Natural Water

1. Answers will vary.

2. Answers will vary.

3. Answers will vary. See the chart on page 11 explaining how acid rain affects animals living in the water.

4. The rainwater should be more acidic than the surface water from the pond or stream. This is because most surface water travels through soil and is buffered, or made less acidic, before reaching a pond or stream. It is important to note that runoff from agricultural areas containing pesticides or fertilizers or runoff of industrial pollutants discharged (legally or illegally) into streams may affect the results of this experiment.

5. Answers will vary.

Experiment 5: Measuring Soil pH

1. Some types of plants and animals are able to live in acid soils, while others are not. Be aware, however, that many factors, not just the soil acidity, determine the types of plants and animals that occur at a particular site.

2. Answers will vary.

3. Some plants require acid soils to grow and thrive. For example, pine trees,

azaleas, rhododendrons, cranberries, blueberries, potatoes, and tomatoes prefer acid soils. However, most plants thrive only in soils of pH 6 to 7.

4. Some soils, such as in many Midwestern states, contain a lot of limestone and are alkaline. In those locations, people often add sulfate, such as ammonium bisulfate to soil to make it less basic.

5. Answers will vary, depending on where the soil samples came from. The underlying rock can have a large effect on soil pH, and so can industrial or agricultural runoff in the area.

6. Answers will vary.

7. Answers will vary.

Experiment 6: Soil Buffering

1. If the pH stayed the same, the soil did not buffer the acid. Each pH value above 4 indicates that the soil buffered increasing amounts of the acid.

2. Not all plants and animals can live in acidic soil, so by buffering the acid the soil makes the environment more habitable.

3. Even soil capable of buffering acids can be overpowered if enough acid is added. As more acid is added to the soil by acid rain, the buffering capability decreases, and the soil becomes more acidic. This can hurt delicate ecosystems like forests.

4. Answers will vary.

Experiment 7: Observing The Influence Of Acid Rain On Plant Growth

1. The plants watered with acidic solutions will not do as well as the plants watered with clean water. Some effects of the acidic solution on plant growth are withering or drooping leaves, change in leaf color, and stunted growth.

2. The effects of the very strong acidic solution on the plants should be visible within a few days. The less acidic solution may not affect the plant for a week to a week and a half. The degree to which the plant is affected by the acid depends on the health of the plant and the hardiness of the species.

3. The plant watered with the very acidic solution should exhibit significantly more damage than the plant watered with the weaker acidic solution.

4. Answers will vary.

Experiment 8: Looking At Acid's Effects On Metals

1. There should be no change.

2. The liquid should be bluish-green. The bluish-green substance in the vinegar comes from the copper in the penny. It is a byproduct of the chemical reaction in which the acid in the vinegar gradually eats away the penny.

3. The chemical reaction between the acid and the copper penny is so slow that you cannot see any difference in the shape of the metal in just 5 days, at least not with your eye alone. You may see some changes after about 2 weeks, especially at the edge of the penny.

4. Answers will vary.

Experiment 9: Observing The Influence Of Acid Rain On Marble And Limestone

1. The chalk in the vinegar should be significantly more worn away then the piece of chalk in the distilled water. This is because the acid dissolves the calcium carbonate in the chalk.

2. Answers will vary.

3. Answers will vary.

4. Answers will vary.

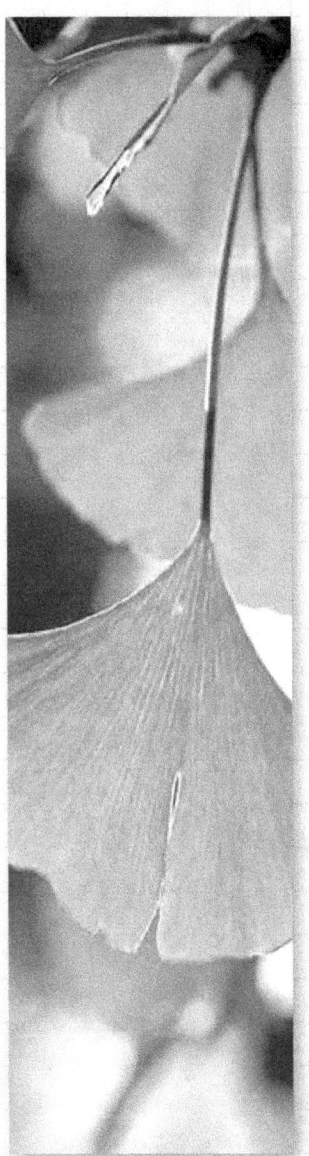

Activities

- Join GLOBE (Global Learning and Observations to Benefit the Environment) at www.globe.gov. Collect data on the pH of rain, surface water, and soil according to GLOBE's protocols and post your results on GLOBE's online database. Compare your results to other student's results in the United States and around the world.

- Role play different characters involved in or affected by acid rain. Each person in the class takes the role of an interested party (for example a fish, bird, coal miner, farmer, factory owner, power company, stream, lake, tree, etc.). With each student or group playing a character, have a group discussion or debate on acid rain. Tell the rest of the class how acid rain affects your character, and then present an argument for or against laws that control acid rain.

- As individuals, contact your local power company. If you have a large class, you may also want to assign some students to research power companies from other parts of the country. Many power companies use more than one source of power to make enough electricity for the community. Some also buy electricity from other power companies. Ask the power company what sources of energy it uses (hydroelectric, nuclear, gas, oil, coal, other) and what percentage of their energy output comes from each source. Write down results to report to the class.

- In small groups or as individuals, write, produce, and direct a special segment for a TV weather special on the effect of weather patterns on the travel of air pollution over large distances. Contact the weather bureau or a local TV station's weather department to ask about the wind patterns in your area. Use the EPA Clean Air Market Division's data and maps Web site (http://camddata andmaps.epa.gov/gdm/) to find power plants and their emissions in your area. The EPA's C-MAP Web site (www.epa.gov/airmarkets/maps /c-map.html) is another online tool that you can use to perform this activity. Map where you think pollution from local power plants may end up. Discuss what communities, cities, ecosystems, parks, schools, etc. may be affected.

Here are some suggested interdisciplinary activities, projects, and field trips to enhance students' understanding of the sources and effects of acid rain.

This power plant in Tennessee uses hydroelectric energy to produce electricity.

Contact your local water company to find out where your water comes from.

- Learn how EPA regulates the emissions that cause acid rain. EPA has several programs that control emissions. Research the Acid Rain Program, NO_x State Implementation Plan (SIP) Call, NO_x Budget Trading Program, or the Clean Air Interstate Rule (www.epa.gov/airmarkets). Write a report on what pollutants these programs regulate and how they regulate them.

- Invite a local natural resource specialist from a zoo, aquarium, nature center, or national/state park to come in and speak to your class about the effects of acid rain on animals and their habitat. You could also telephone natural resource specialists around the country. Interview the specialist and ask them to tell you about the impact of acid rain and dry deposition on the lakes, forests, or other natural resources in your area. Write down what you have learned in a report.

- EPA has regional offices around the nation. Contact your regional EPA office, and ask a representative from EPA to visit your class. Prepare for their visit by brainstorming interview questions to ask them when they visit. You may want to ask them about the effects of acid rain in your region of the country, or what other environmental issues they believe are most pressing in your region. When they come to class, take turns asking questions, and write a newspaper article about the interview. The locations of EPA regional offices and contact information can be found online at www.epa.gov /epahome/locate2.htm.

- Call or visit your local water company. Ask them where your water comes from: a well, lake, reservoir, or river. If you have a private well, ask your parents if the water is treated for acidity, and if so, how. When talking to the water company, ask how they treat the water for acidity. Ask if they can tell you the pH of the water before it is treated, and after it is treated. Is it completely neutralized? Write down their answers in a report to give to the class.

- Contact a local architecture firm, an architecture department at a local university, a member of a state or national Green Building Council, or the EPA ENERGY STAR® program, and see if you can get someone to visit your classroom and describe how homes, schools, and office buildings can use energy and resources more efficiently.

Contact a local natural resource specialist and have them come in to speak about the effects of acid rain on animals and their habitat.

Games

Crossword Puzzle

DIRECTIONS

Read each clue below then find the number in the puzzle that corresponds to each clue. To assist you, there is a list of possible answers to each clue below the puzzle. Now it is time to see how many clues you can get correct!

ACROSS

3. A solution is _____ when it has a pH higher than 7.0.

5. When power plants burn _____, they release sulfur dioxide and nitrogen oxides into the air.

7. Wet _____ refers to acidic rain, fog, and snow.

9. _____ deposition can be wet or dry.

11. You can _____ acidic water by adding a base.

13. A _____ removes sulfur dioxide from the gases leaving the smoke-stack of a power plant.

DOWN

1. The government gives an _____ to a power plant, letting it release a set amount of sulfur dioxide.

2. One way that people can help prevent acid rain is by joining a _____, in which individuals share rides to their destination and reduce the number of cars polluting the air.

4. _____ is to turn from gas or vapor into liquid form.

6. An _____ consists of plants and animals and the environment in which they live.

8. Sulfur dioxide, nitrogen oxides, ozone, and particulate matter are examples of _____.

10. Hydroelectricity is produced from the energy of running _____.

12. Solar energy is energy that comes from the _____.

POSSIBLE ANSWERS

Acid	Ecosystem
Allowance	Neutralize
Basic	Pollutants
Carpool	Scrubber
Coal	Sun
Condense	Water
Deposition	

Word Search

DIRECTIONS

All words that are listed below are hidden in the collections of scrambled letters. Your goal as an acid rain detective is to find as many words as possible. Look very carefully — the words can go forwards, backwards, up and down, and diagonally. How good are your detective skills?

FIND WORDS

Acidic

Allowance

Basic

Buffer

Cap

Carpool

Chemical

Coal

Deposition

Ecosystem

Emissions

Geothermal

Hydropower

Pollution

Scrubbers

O	A	P	Q	O	L	A	C	I	M	E	H	C	P	J	U	J
I	N	A	N	X	F	M	U	N	J	Q	H	V	G	I	H	G
J	L	M	S	O	Z	H	I	B	A	S	I	C	E	X	Y	C
D	S	L	E	N	I	A	C	L	I	H	R	H	O	N	D	T
E	N	C	X	T	Q	T	Q	Z	V	D	A	J	T	P	R	N
C	O	O	R	M	S	R	A	H	F	X	T	N	H	D	O	O
N	I	P	D	U	Z	Y	V	C	V	M	B	T	E	A	P	I
A	T	B	U	F	B	Y	S	C	I	U	Q	C	R	A	O	T
W	I	S	R	Y	W	B	Y	O	F	F	L	S	M	C	W	U
O	S	R	D	Y	R	B	E	F	C	X	I	J	A	I	E	L
L	O	S	X	Y	O	L	E	R	G	E	X	D	L	D	R	L
L	P	H	P	T	O	R	R	O	S	V	T	A	I	I	P	O
A	E	S	N	O	I	S	S	I	M	E	W	P	P	C	D	P
S	D	N	P	L	E	A	C	H	I	N	G	Q	L	T	A	G
Z	C	R	Y	T	A	Z	H	X	M	N	G	X	I	U	W	O
F	A	H	P	X	R	A	L	N	Z	W	C	A	P	E	X	X
C	I	A	M	C	O	A	L	H	Q	G	Q	D	D	B	O	K

Game Answers

Crossword (Answers):

- 3/1 Across: BASIC
- 5 Across: COAL
- 7 Across: DEPOSITION
- 9 Across: ACID
- 11 Across: NEUTRALIZE
- 13 Across: SCRUBBER
- 1 Down: ALLOWANCE
- 2 Down: CARPOOL
- 3 Down: B
- 4 Down: CONDENSE
- 6 Down: ECOSYSTEM
- 8 Down: POLLUTANT
- 10 Down: WATER
- 11 Down: NM (NITROGEN)
- 12 Down: SN (SOOT)

Word Search (Answers):

```
O A P Q O L A C I M E H C P J U J
I N A N X F M U N J Q H V G I H G
J L M S O Z H I B A S I C E X Y C T
D S L E N I A C L I H R H O N D R T
E N C X T Q T Q Z V D A J T P R O N
C O O R M S R A H F X T N H E A O I
N I P D U Z Y V C V M B T E R A O T
A T B U F B Y S C I U Q C R A O U U
W I S R Y W B Y O F F L S M C W L L
O S R D Y R B E F C X I J A I D E L
L O S X Y O L E R G E X D L I D R O
L P H P T O R R O S V T A I I P P
A E S N O I S S I M E W P P C D P
S D N P L E A C H I N G Q L T A G
Z C R Y T A Z H X M N G X I U W O
F A H P X R A L N Z W C A P E X X
C I A M C O A L H Q G Q D D B O K
```

Additional Resources

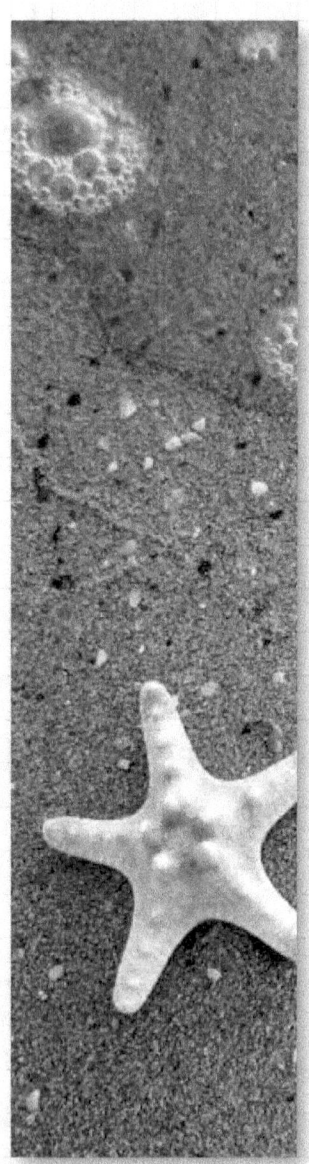

Readings

Parks, Peggy J. *Acid Rain (Our Environment Series)*. Detroit [MI]: KidHaven Press, 2006. Ages 12 and up.

Petheram, Louise. *Acid Rain*. Mankato [MN]: Capstone Press, 2002. Ages 8 to 9.

O'Connor, Rebecca K. *Acid Rain (Lucent Overview Series)*. Farmington Hills [MI]: Lucent Books, 2003. Ages 12 and up.

Allaby, Michael. *Fog, Smog, and Poisoned Rain (Facts on File Dangerous Weather Series)*. New York [NY]: Facts on File, May 2003. Ages 11 and up.

Johnson, Rebecca L. *Acids and Bases (Physical Science Series)*. New York: National Geographic, 2004. Ages 9 through 13.

Web sites

EPA has information on acid rain and other air quality issues at www.epa.gov /airmarkets/index.html.

Find resources and activities on EPA's general site for teachers and students at www.epa.gov/region5/teachers.

The EPA's Acid Rain for Kids site can be accessed at www.epa.gov/acidrain /education/site_kids.

The Acid Rain Program Annual Reports are available online at www.epa.gov/airmarkets /progress/progress-reports.html.

The U.S. Geological Survey Web site has an acid rain page at http://ga.water.usgs .gov/edu/acidrain.html.

The National Park Service has a great students and teachers Web site at http:// www2.nature.nps.gov/air/edu/index.cfm.

The Geological Society of America has information on several environmental issues at www.geosociety.org /educate /LessonPlans/i_water.htm.

The National Atmospheric Deposition Program site has some more information for teachers or older students, including links to data and maps at http://nadp .sws.uiuc.edu/. It also has a page that links to several good online resources for all ages at http://nadp.sws.uiuc.edu /cal/Educational_Information.htm.

The National Oceanic and Atmospheric Administration has a students and teachers resource page at http://oceanservice.noaa.gov/kids.

You can look at acid rain data collected by students around the world, or even upload your own on the GLOBE Web site at www.globe.gov/fsl/welcome.html.

Audiovisuals

Acid Rain (DVD) Educational Video Network, Inc., 2004. Run Time: 20 minutes.

Air Pollution, Smog, and Acid Rain (DVD) Educational Video Network, Inc., 2004. Run Time: 21 minutes.

Glossary

Acid Any of a large group of chemicals with a pH less than 7. Examples are battery acid, lemon juice, and vinegar.

Acidic Describes a substance with a pH less than 7.

Acid Deposition Acidic material that falls from the atmosphere to the Earth in either wet (rain, sleet, snow, fog) or dry (gases, particles) forms.

Acid Rain Rain that has become acidic by contact with air pollution. Other forms of precipitation, such as snow and fog, are also often included in the term acid rain or acid wet deposition.

Allowance The permission, given by the government, to emit a certain amount of sulfur dioxide (SO_2) or nitrogen oxide (NO_x).

Aluminum A silvery-white metal; it is the most abundant metal in the Earth's crust.

Aquatic Relating to water.

Atmosphere The air or gases that surround a planetary body such as the Earth.

Atmospheric Relating to the atmosphere, or the air above the Earth.

Base Any of a large group of chemicals with a pH greater than 7. Examples are ammonia and baking soda.

Basic Describes a substance with a pH greater than 7. Another word for basic is alkaline.

Buffer A substance, such as soil, bedrock, or water, capable of neutralizing either acids or bases.

Buffering Capacity The ability of a substance to resist changes in pH when acids or bases are added.

Cap A national limit that is placed on the amount of a pollutant that can be emitted. The cap is very important because it makes sure that emissions of a pollutant are reduced.

Cap and Trade An environmental policy tool that controls large amounts of emissions from a group of sources. Cap and trade programs set a cap, or limit, on emissions. Then allowances for emissions are traded between sources, so that economic market forces allow large emissions reductions to be cost-effective.

Carbon Dioxide (CO_2) A naturally occurring gas made carbon and oxygen. Sources of carbon dioxide in the atmosphere include animals, which exhale carbon dioxide, and the burning of fossil fuels and biomass.

Condense To change from gas or vapor to liquid form.

Deposition When chemicals like acids or bases fall to the Earth's surface. Deposition can be wet (rain, sleet, snow, fog) or dry (gases, particles).

Dry Deposition The falling of small particles and gases to the Earth without rain or snow.

Ecology The study of ecosystems. Someone who studies ecology is called an ecologist.

Ecosystem All the living and nonliving things in an area, as well as the interactions between them.

Emissions The gases that are released when fossil fuels are burned.

Energy Resources Natural resources that can be used to make heat, electricity, or any other form of energy. The most commonly used energy resources are fossil fuels (coal, oil, and gas), but the sun, wind, and anything else that makes energy are also energy resources.

Environment The air, water, soil, minerals, organisms, and all other factors surrounding and affecting an organism.

Evaporate To change from liquid into gas.

Fossil Fuels Oil, natural gas, and coal. Fossil fuels were made in nature from ancient plants and animals, and today we burn them to make energy.

Fuel Cells Similar to batteries, fuel cells store energy that can be used to power all sorts of things. Unlike a battery though, fuel cells do not "run out" and do not need to be recharged or replaced.

Greenhouse Gases Gases that occur naturally in the Earth's atmosphere and trap heat to keep the planet warm. Some examples are carbon dioxide, water vapor, halogenated fluorocarbons, methane, hydrofluorocarbons, nitrous oxide, perfluoronated carbons, and ozone. Some human actions, like the burning of fossil fuels, also produce greenhouse gases.

Habitat The place where a plant or animal lives and grows, such as a forest, lake, or stream.

Haze When particles of dust, pollen, or pollution make the air less clear, and limit visibility.

Hydroelectric Power Energy that is generated by dams, which use water to turn turbines and generate electricity.

Hydrologic Cycle The movement of water from the atmosphere to the surface of the land, soil, and plants and back again to the atmosphere.

Litmus Paper Paper coated with a chemical coloring obtained from lichens that turns red in acidic water and blue in basic water. It is used as an acid-base indicator.

Meteorological Pertaining to the weather.

Natural Resources All the parts of the Earth that are not human-made and which people use, like fish, trees, minerals, lakes, or rivers.

Neutral A substance that is neither an acid nor a base and has a pH of 7. Neutral substances can be created by combining acids and bases.

Neutralize To combine acids and bases to make a neutral substance or solution. For example, acidic water can be neutralized by adding a base.

Nitric Acid An acid that can be produced from nitrogen oxide, a pollutant that results from the burning of fossil fuels.

Nitrogen Oxides (NO_x) A family of gases made up of nitrogen and oxygen commonly released by burning fossil fuels.

Nuclear Power Energy that comes from breaking apart the center (nucleus) of an atom.

Nutrient Deficiency When a living thing lacks the vitamins and minerals it needs to survive.

Ozone A chemical that is made of three oxygen atoms joined together, and found in the Earth's atmosphere. There are two kinds of ozone: good ozone, and bad ozone. Good ozone is found high in the Earth's atmosphere, and prevents the sun's harmful rays from reaching the Earth. Bad ozone is found low to the ground, and can be harmful to animals and humans because it damages our lungs, sometimes making it difficult to breathe.

Ozone Layer The layer of ozone that shields the Earth from the sun's harmful rays.

Particulate Matter Tiny solid particles or liquid droplets suspended in the air.

pH Paper Paper that changes color to show the pH of a substance.

pH Scale The range of units that indicate whether a substance is acidic, basic, or neutral. The pH scale ranges from 0 to 14.

Photosynthesis The process that plants use to convert sunlight to energy to live and grow.

Pollutants Chemicals or other substances that are harmful to or unwanted in the environment. Some examples of pollutants are sulfur dioxide (SO_2), nitrogen oxides (NO_x), ozone, and particulate matter.

Pollution The release of harmful substances into the environment.

Precipitation Water falling to the Earth. Mist, sleet, rain, hail, and snow are the most common kinds of precipitation.

Primary Producers Organisms that use photosynthesis to produce their own food. All plants are primary producers. Primary producers are the base of the food chain because they feed everything else.

Reactive Having the tendency to chemically combine with something else and change its form. For example, a strong acid is highly reactive with a strong base.

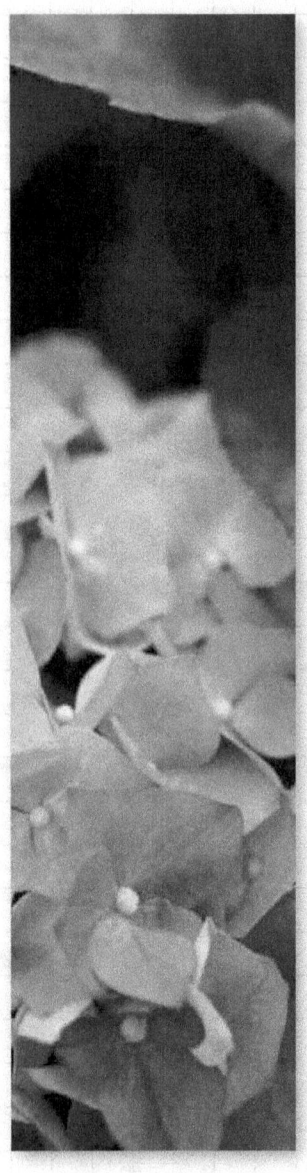

Respiratory Illness Diseases affecting the organs we use to breathe. Asthma, bronchitis, and pneumonia are examples of respiratory illnesses.

Respiratory System The organs in our body involved with the process of breathing.

Runoff Water that flows off land into lakes and streams.

Scrubber A device that removes air pollution, mainly sulfur dioxide, from smokestacks.

Solar Power Electricity that is generated by harnessing the energy of the sun. Solar panels are often used to convert sunlight to energy.

Sulfur Dioxide (SO_2) A naturally occurring gas made of sulfur and oxygen that is also released when fossil fuels are burned.

Sulfuric Acid An acid that can be produced in the atmosphere from sulfur dioxide, a pollutant that results from burning fossil fuels.

Surface Water Water that is found on the surface of the Earth in lakes, ponds, streams, or rivers.

Turbine A motor activated by water, steam, or air to produce energy.

Wind Power Energy that is generated when the wind turns the sails of a windmill, which are attached to turbines that generate electricity.

This certificate is awarded to

For participation and completion of

The Acid Rain Awareness Program

Teacher Signature

Date

Certificate of Achievement